U0161518

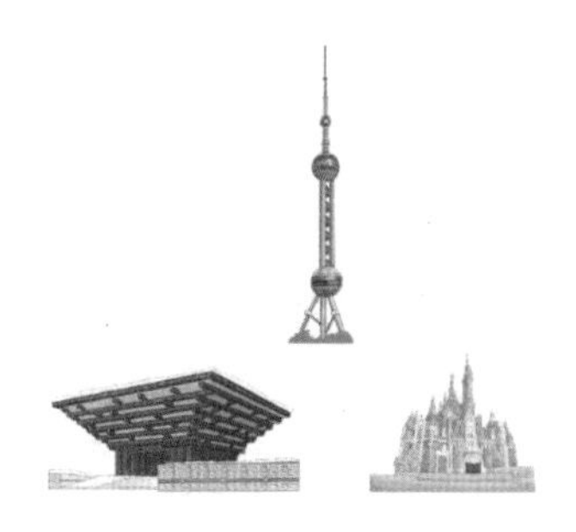

Chinese Cities
Shanghai Impressions

中国城市
上海印象

五洲传播出版社
China Intercontinental Press

Shanghai-style
海派
风花雪月

New Culture
文艺
时尚新生活

24 廿四
新新大旅馆

Indulging in Nostalgia

怀旧 百年故事

The Bund

Story of Shanghai

外滩 | 上 海 的 故 事

At the end of the 19th Century, foreign consulates and banks built Western style houses along the Huangpu River. In the early 20th Century, the Bund, known as the "Far East Wall Street" appeared at the end of Nanjing Road, becoming a witness to the vicissitudes of Shanghai.

19世纪末，外国领事馆和银行纷纷在黄浦江边盖起洋房，不同国家设计风格的建筑聚集在这里。20世纪初，被称为"远东华尔街"的外滩位于南京路尽头，记录下上海曾经的沧桑。

Nanjing Road

Between Tradition and Modernity

南京路｜传统与现代

Nanjing Road Pedestrian Street is a microcosm of the interweaving of old and new Shanghai. The four department stores in Shanghai are all located here, giving a new look to the development of the area. The mix of tradition and modernity adds additional another charm to the old street today.

南京路步行街是新旧上海交织的缩影。曾经的老上海四大百货商店都在这里，把南京路的过去用新的面貌呈现出来。传统与现代的交织为这条百年老街增添了别样的魅力。

Fuzhou Road

A Street of Culture and Literature

福州路 | 四马路文化

This quiet and simple road is the source of Shanghai culture. In 1843, Mohai Bookstore, first of its kind in China, was founded on Fuzhou Road. Scholars proficient in Chinese and Western culture were gradually gathered here, and became the primary creators of the distinctive Shanghai culture. This road forms the cultural source linking old and new Shanghai for visitors.

这条静默而朴素的马路是海派文化的源头。1843年，中国第一个出版机构墨海书馆在福州路创立。这里逐渐汇聚起精通中西文化的文人，他们便是最初的海派文化始作俑者。这条马路正向步履匆匆的行人诉说着老上海的文化源头与过往。

North Sichuan Road
A Glimpse of Old Shanghai's Prosperity

四川北路 | 老上海的热闹

This is the first road built after Shanghai opened to the outside world. The Yingzi Photo Studio, Xinhua Bookstore, Sixin Snacks Shop and No.7 Department Store opened one after the other. Marks of the lives of many old Shanghai people remain. It is this part of Shanghai where many took their first pictures, spent Christmas for the first time and tried their first Western dishes.

四川北路是上海开埠后最早建成的马路，英姿照相馆、新华书店、四新点心店、第七百货纷纷在这里落户。这里留下太多老上海市民的生活印记，很多人在这里尝试了人生第一次拍照、第一次过圣诞节、第一次西餐。

Revisiting the Past

访古 老城记忆

The old City God Temple

Memory of the Old City

老城隍庙 | 老城厢的记忆

The old City God Temple is a focal point for the first memories of many Shanghai people. The temple occupies a very sacred position in their hearts. Every holiday, many come to piously offer a wish before the City God. Curls of smoke waft into the air, carrying people's hopes for the future.

老城隍庙是上海人对老城厢生活的最初记忆。城隍庙在老上海人的心里有着很神圣的地位，每到节假日，大家都会迈着虔诚的步伐去城隍前许个愿。袅袅的青烟在老城厢里飘来荡去，带着人们对未来的美好希望。

Yuyuan Garden

Garden Unique to Southern Yangtze River Area

豫园 | 古朴的江南园林

Yuyuan Garden, with a history of more than 400 years, features a style unique to the Ming and Qing Dynasties (1368-1911). A wall painted with flying dragons divides the garden into different scenic spots, revealing the rich scenery of traditional Chinese gardens. These include pavilions, rockeries and ponds.

豫园距今已经有 400 多年的历史了，具有明清两代南方建筑艺术风格。豫园的围墙上，游龙蜿蜒起伏，将豫园分隔成不同景区，透出园林丰富的景层。园内亭台楼阁、假山池沼，景致各别，有“以小见大”的江南园林特色。

Longhua Temple

The Bell Tolls for New Year's Eve

龙华寺 | 敲响新年的晚钟

The is the oldest and largest ancient temple near Shanghai. It is most famous for the Millennium Pagoda, Longhua Temple Fair and Longhua Evening Bell.

龙华寺是上海附近历史最悠久、规模最大的古刹。这座古刹以千年古塔、龙华庙会和龙华晚钟最为出名。

Zhujiajiao
Venice of Shanghai

朱家角 | 上 海 威 尼 斯

This is the most well-preserved ancient town in Shanghai. In history, many merchants would gather here. Today, there are ancient buildings in Ming and Qing styles. The zigzag stone-paved path twists and turns, linking various streets. Small bridges span a small river on which oarsmen propel their small boats. The banks are lined with weeping willows creating an almost poetic scene.

朱家角是上海保存最完好的江南水乡古镇。历史上曾经商贾云集。如今，这里随处可见古色古香的明清时期的老建筑，弯弯曲曲的石板小径在街巷间迂回曲折。小桥流水、乌篷小船、岸边杨柳，往昔的生活场景依稀可见。

茶

Renowned Figures

名人 岁月的山河

Duolun Road

Memory of modern history of Shanghai

多伦路｜上海近代历史的记忆

The L-shaped Duolun Road was once an important stronghold of modern Chinese literature. Today, it is still lined with houses of different styles, some of which are the residences of cultural celebrities such as Lu Xun, Mao Dun, Ye Shengtao and Roushi, whose art prospered in the 1930s. It is not within a foreign concession, but has a close connection. Western culture is also highlighted here. Lu Xun used to talk about the national and international situation with friends in local cafes and bookstores.

呈“L”状的多伦路曾经是中国近代文学的重要据点。如今多伦路的街道两旁，风格各异的建筑依次而立。这些石库门房子是 20 世纪 30 年代鲁迅、茅盾、叶圣陶、柔石等一大批文化名人的住所。这里既非租界，又跟租界丝缕相连，西洋文化气息恰到好处。当年鲁迅就经常在咖啡馆和书店里边喝咖啡边谈论时局。

The Former Residence of Lu Xun in Shanghai

Splendor in a Humble Dwelling

鲁迅故居 | 弄堂里的辉煌历史

No.9 Lane 132 on the Shanyin Road is the last place where Lu Xun resided in Shanghai. Now, except for the words marking this as Lu Xun's Former Residence, everything in this small building is plain. However, it is here he wrote nine miscellaneous works and *Old Retold* and edited a series of cultural journals such as *Yusi and Zhaohua*, until he passed away on October 19, 1936.

山阴路 132 弄 9 号是鲁迅先生在上海居住的最后一个地方，也是他人生的最后一个客栈。如今，这座小楼除了鲁迅故居那几个字有些显眼外，一切都显得平淡无奇。然而，先生正是在这里留下了最辉煌的作品。在这里，鲁迅出版了 9 本杂文集和《故事新编》，编辑了《语丝》《朝花》等系列文化期刊，直到 1936 年 10 月 19 日，鲁迅先生与世长辞。

Former Residence of Sun Yat-Sen

Still time

孙中山故居 | 静止的时光

Sun Yat-sen's Former Residence is located on the western side of Fuxing Park, a quiet European-style garden. In front of the building is a square lawn surrounded by holly, magnolia, camphor and pine trees. Sun's bronze statue looks out over today's bustling surroundings.

孙中山故居位于复兴公园西侧，是一座幽静的欧式花园。楼前是一片正方形的草坪，三面环绕着冬青、玉兰、香樟和松柏。先生的铜塑像日复一日地静静坐在那里，看着今日的市井尘烟。

Moller Villa

Little girl's dream

马勒别墅 | 小女孩的梦

In 1926, a little girl in Shanghai dreamt she lived in Hans Christian Andersen's fairy tale castle. The little girl's father built this Mahler Villa according to this daughter's dream to the great surprise of the whole of Shanghai. From a private house in earliest days it has now become a Mahler Villa Hotel. From spring to autumn, as the clouds come and go, it always looks the same – like something out of a fairy tale.

1926 年的上海，一个小女孩做了一个奇特的梦，梦到自己住在了安徒生的童话城堡里。小女孩的爸爸按照女儿梦中的情景建造了这座童话城堡——马勒别墅。这是一座令整个上海惊奇的建筑，从最早的私人住宅到如今成为马勒别墅饭店，春去秋来，云起云落，而它始终如同从童话中走来一样，容颜不改。

Shanghai-style

海派 风花雪月

Sinan Road
The Most Beautiful Road in Shanghai

思南路 | 上海最美的马路

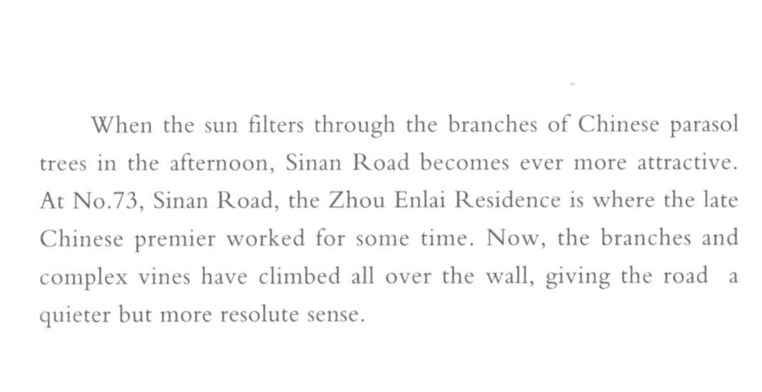

When the sun filters through the branches of Chinese parasol trees in the afternoon, Sinan Road becomes ever more attractive. At No.73, Sinan Road, the Zhou Enlai Residence is where the late Chinese premier worked for some time. Now, the branches and complex vines have climbed all over the wall, giving the road a quieter but more resolute sense.

思南路是一条安静的马路，下午阳光从梧桐树的枝杈倾泻下来时，思南路就愈发迷人。思南路的 73 号周公馆是当年周恩来在上海工作的地方，如今枝条繁复的藤蔓已经爬满整个墙壁，思南路愈发多了一份沉静与坚毅。

Calendar Posters

Classic Charm of Shanghai

月份牌 | 海派的花样年华

Shanghai was once known as the Oriental Paris, which means it was once the most Westernized city in China, pursuing the most elegant and exquisite lifestyle. In 1896, Calendar Posters came to Shanghai as a commercial advertisement. Thirty years later, they were considered the height of fashion and a must-have decoration in the city. The new women wearing a cheongsam featured in these posters have become a symbol of gentle and elegant life of Shanghai.

上海曾经被称为东方的巴黎，曾经是中国最西化的城市，追求最优雅精致的生活方式。1896 年，作为商品广告的月份牌来到上海。30 年后月份牌蔚为风尚，成为上海家家必有的装饰。衣着旗袍的新派的女子月份牌也就成为了吴侬软语的海派生活标志性老物件。

Shanghai Cuisine

The taste of old Shanghai

本帮菜 | 菜浓油赤酱的记忆

The most basic flavor of Shanghai common dishes feature thick oil and dark soy sauce, which is the original flavor of the dishes. Later, Shanghai cuisine, improved by hybridization, took the advantages provided by many families and formed a school of its own.

上海百姓家常菜最基本的味道就是“浓油赤酱”，是本帮菜的原味。后来，经过杂合改良出的上海菜则博取众家之长，自成一派。

Paramount Ballroom

The Beholder of the Distant Past

百乐门 | 往事依稀

Mr. Pai Hsien-yung's *The Last Night of Madam Chin* made Paramount famous again after its period of decline. Paramount was once the first Music House in the Far East, with numerous legends. Chin Manli, the former dancer, was shot in the ballroom because she didn't want to dance with a Japanese. When Liang Shiqiu's wife, Han Jingjing, at the age of 11, she stood out among 3,000 aspiring candidates and became a singer. Now, the former bustling place has become a silent old building on Yuyuan Road.

白先勇先生的《金大班的最后一夜》让百乐门在没落之后又一次声名远扬。百乐门是曾经的远东第一乐府，发生过无数传奇故事。当年的舞女陈曼丽因为不愿为日本人伴舞而被枪杀在舞厅内。梁实秋的妻子韩菁菁 11 岁的时候在 3000 名考生中脱颖而出，成为一代歌后。星转斗移、沧海桑田，昔日的繁华场如今已成为愚园路上一处静默的老建筑。

New Culture

文艺 时尚新生活

The Oriental Pearl Tower and Pudong New Area

The highland of Shanghai

东方明珠与浦东新区 | 上 海 的 高 地

Pudong is a brand new area represented by Lujiazui. The Huangpu River empties into the sea at this point. At night, the bright neon lights create the most beautiful city night scene in China.

浦东是一个崭新的区域，而陆家嘴就是浦东的代表。风和日丽时，黄浦江上巨轮如梭，连绵入海。夜晚，光芒四射的霓虹灯点亮了上海的夜空，点亮了中国最美的不夜城。

Shikumen and Xintiandi

Shanghai's tradition and fashion

石库门与新天地 | 传统与时尚

Lanes in the Shikumen area developed in the 1860s. It was the living area of ordinary families in Shanghai, dwarfed by emerging Xintiandi. The green brick walkway, the red and green clear water brick wall, the thick black paint gate and the lintel carved with Baroque scroll-like mountain flowers make people feel like they are back in old Shanghai.

石库门一带的里弄在 19 世纪 60 年代兴起。石库门是上海普通人家的生活场，因为新天地的出现成为了时髦的代名词。青砖步行道、红青相间的清水砖墙、厚重的乌漆大门以及雕着巴洛克卷涡状山花的门楣，让人们仿佛又一次置身于往日的上海滩。

Fudan University was founded in 1905, adopting the idea of "the morning sun rising again" pinning hopes on the intellectuals running the school independently. Fudan has become a beautiful memory for many with today's sunny campus with gentle summer grass and breezes, from public school to private university, and then from private university to national university.

创建于 1905 年的复旦大学取意于“日月光华、旦复旦兮”，寄托了当时知识分子自主办学、教育强国的希望。百年来，复旦从公学到私立大学，再从私立大学到国立大学，几经辗转，成为今日夏草轻轻、微风吹拂、阳光明媚的校园的一段记忆。

Disney

Fairytale World

迪士尼 | 童话的世界

Every child has a fairy tale world in their heart. Mickey Mouse, Donald Duck, Snow White and the Seven Dwarfs have left indelible memories on the minds of many children. Disney has come to Shanghai and satisfied everyone's fairy tale dreams.

每个孩子的心中都有一个属于自己的童话世界。米老鼠、唐老鸭、白雪公主与小矮人成为了很多小孩子的童年记忆。迪士尼来到上海，满足了每一个人的童话梦。

图书在版编目（C I P）数据

上海印象：汉英对照 / 达雅著. -- 北京：
五洲传播出版社，2020.6
（中国城市）
ISBN 978-7-5085-4444-1

Ⅰ.①上… Ⅱ.①达… Ⅲ.①本册②上海－概况－汉、英
Ⅳ.① TS951.5 ② K295.1

中国版本图书馆 CIP 数据核字 (2020) 第 074856 号

中国城市：上海印象
Chinese Cities：Shanghai Impressions

出 版 人：荆孝敏
责任编辑：杨 雪
设计策划：青芒时代
插　　画：王建华　王梦云　李　进
文　　字：达　雅
译　　者：王国振
装　　帧：张伯阳
出版发行：五洲传播出版社
地　　址：北京市海淀区北三环中路 31 号生产力大楼 B 座 6 层
邮　　编：100088
发行电话：010-82005927，010-82007837
网　　址：http://www.cicc.org.cn，http://www.thatsbooks.com
印　　刷：北京顶佳世纪印刷有限公司
版　　次：2020 年 7 月第 1 版第 1 次
I S B N：978-7-5085-4444-1
开　　本：787mm × 1092mm　1/32
印　　张：6
字　　数：20 千
定　　价：49.8 元